THE POETRY OF SODIUM

The Poetry of Sodium

Walter the Educator™

SKB

Silent King Books a WhichHead Imprint

Copyright © 2023 by Walter the Educator™

All rights reserved. No part of this book may be reproduced in any manner whatsoever without written permission except in the case of brief quotations embodied in critical articles and reviews.

First Printing, 2023

Disclaimer
This book is a literary work; poems are not about specific persons, locations, situations, and/or circumstances unless mentioned in a historical context. This book is for entertainment and informational purposes only. The author and publisher offer this information without warranties expressed or implied. No matter the grounds, neither the author nor the publisher will be accountable for any losses, injuries, or other damages caused by the reader's use of this book. The use of this book acknowledges an understanding and acceptance of this disclaimer.

Chemical Element Poetry Book Series
by Walter the Educator™
"Earning a degree in chemistry changed my life!"
— Walter the Educator

dedicated to all the chemistry lovers, like myself, across the world

CONTENTS

Dedication v

Why I Created This Book? 1

One - Brilliance Illuminates 2

Two - Taste Of The Sea 4

Three - Earth's Embrace 6

Four - Mesmerizing Element 8

Five - Our Praise 10

Six - Style And Grace 11

Seven - Essential To Life 13

Eight - Sodium Whispers 14

Nine - Spark Of Creation 16

Ten - Grand Decree 18

Eleven - Igniting Our Curiosity 20

Twelve - Glorious Dream 22

Thirteen - Flash And A Sizzle 24

Fourteen - No Hesitation 26

Fifteen - Chemical Waltz 28

Sixteen - Rare And Fine 30

Seventeen - Speeding Up Processes 32

Eighteen - Fiery Art 34

Nineteen - Silent Hero 36

Twenty - Compounds And Mixtures 38

Twenty-One - Regulating Balance 40

Twenty-Two - Salty Waves 42

Twenty-Three - Pole To Pole 44

Twenty-Four - Vigor And Might 46

Twenty-Five - Symphony Of Life 47

Twenty-Six - Presence Felt 49

Twenty-Seven - The Catalyst 51

Twenty-Eight - Sodium Revered 53

Twenty-Nine - Catalyst Supreme 55

Thirty - Ever True 57

Thirty-One - New Worlds Emerge 59

Thirty-Two - Enhancing The Taste 61

Thirty-Three - Reigns Supreme 63

Thirty-Four - World Adorns 65

Thirty-Five - Diverse And Grand 67

Thirty-Six - Close To Our Heart 69

About The Author 71

WHY I CREATED THIS BOOK?

Creating a poetry book about the chemical element of Sodium is a unique and creative way to explore the properties and characteristics of this element. By using poetry, I can describe the various uses and applications of Sodium, as well as its role in the natural world. Additionally, poetry can be used to convey the more abstract qualities of Sodium, such as its potential for explosive reactions or its ability to form strong bonds with other elements. Overall, this poetry book about Sodium can provide a fresh and engaging perspective on this important element.

ONE

BRILLIANCE ILLUMINATES

In the realm of elements, shining bright,
Lies a metal that ignites the darkest night.
Sodium, the name that graces the stage,
With atomic number 11, it takes its place.

A dance of electrons, a symphony of fire,
Sodium gleams with an incandescent desire.
Soft and silvery, it reflects the sun's embrace,
A gleaming treasure in the Earth's vast space.

In the depths of oceans, where waves crash and foam,
Sodium resides, a wanderer from its home.
With a fiery temperament, it yearns to be free,
A restless spirit, dancing in the sea.

In compounds it weaves, a magical spell,
Creating salts that many know so well.

From table salt to baking soda's might,
Sodium's presence brings flavor and light.
    In the laboratory, it sparks a reaction,
Exploding with fervor, causing a fraction.
A burst of light, a sizzling sound,
Sodium's energy, forever renowned.
    But tread with caution, for sodium's might,
Can cause a blaze, a blinding light.
Handle with care, lest chaos ensue,
Harness its power, and miracles ensue.
    Oh, sodium, element of fiery glow,
From ancient times, you continue to flow.
A symbol of energy, a catalyst's delight,
Your brilliance illuminates the darkest night.

TWO

TASTE OF THE SEA

In the realm of elements, behold Sodium's might,
A fiery presence, a star in the night.
With brilliance it dazzles, a captivating glow,
A symbol of energy, a story to bestow.

From ancient past to modern day,
Sodium's essence lights the way.
In stars it dances, a celestial fire,
A beacon of hope, a cosmic desire.

In the depths of oceans, where life abounds,
Sodium's ions, harmony it compounds.
In saline tears, a taste of the sea,
Sodium's flavor, a symphony.

But heed my words, handle with care,
For Sodium's temper, beware and be aware.
Its explosive nature, a cautionary tale,
A reminder of power, a caution to prevail.

Yet despite its dangers, Sodium prevails,
A catalyst for progress, where innovation trails.
In laboratories, it sparks creation,
Illuminating pathways, a scientific revelation.
So let us celebrate Sodium's grace,
A radiant element, with a fiery embrace.
From ancient civilizations to modern times,
Sodium's brilliance forever shines.

THREE

EARTH'S EMBRACE

In a realm of fiery light, Sodium gleams,
A metal that dances in vibrant beams.
Its atomic number, a proud eleven,
Unleashing wonders since time was given.

Beneath the surface of a tranquil sea,
Sodium lies, waiting patiently.
With a sizzle and a spark, it takes its form,
A catalyst for life, a chemical norm.

In laboratories, it's a trusted aid,
In reactions, it plays a vital trade.
From soap to glass, it lends its grace,
Enhancing products we embrace.

But caution is required in Sodium's realm,
For it possesses a fiery helm.
With water, it reacts with a fierce display,
Exploding with vigor, lighting the way.

In the deep corners of the Earth's embrace,
Sodium resides, a treasure to trace.
From salt flats to brine lakes, it resides,
A testament to nature's bountiful tides.

So let us celebrate Sodium's might,
A crucial element, shining so bright.
From chemistry's crucible to nature's hold,
Sodium enriches, its story yet untold.

FOUR

MESMERIZING ELEMENT

In the realm of chemistry's embrace,
Lies a metal, Sodium, with grace.
A spark of brilliance in the deep,
An element that stirs passions, we keep.

From ancient times, its flame did glow,
Guiding sailors where the oceans flow.
Its light, a beacon in the night,
A symbol of hope, a source of might.

Within our bodies, Sodium does reside,
Controlling nerves, with every stride.
A messenger, it carries the charge,
In every heartbeat, a rhythm at large.

In compounds, Sodium finds its way,
Creating salts, like a dancer at play.

From table to ocean, it freely moves,
A bond with Chlorine, it sweetly proves.
    In the lamp's flicker, Sodium shines,
Illuminating streets and ancient shrines.
An orange glow, a warm embrace,
A glowing reminder of Sodium's grace.
    From fireworks high up in the sky,
To soda bubbles, sparkling by.
Sodium's touch, a magical sight,
A symphony of colors, a dazzling light.
    In laboratories, Sodium's allure,
A catalyst for reactions, pure.
Its fusion with science, a wondrous art,
Unveiling secrets, unlocking the heart.
    So let us celebrate Sodium's might,
Its presence shining, day and night.
In every aspect, it plays its role,
A mesmerizing element, capturing the soul.

FIVE

OUR PRAISE

In the heart of oceans, deep and vast,
Sodium resides, a treasure unsurpassed.
Its atoms dance, with electrons aglow,
A symphony of light, that only few know.

Through the waves, its power extends,
Creating salts, where life transcends.
A catalyst of taste, a flavor divine,
In every dish, a sprinkle, a sign.

But beware, for Sodium holds a spark,
Explosive nature in the dark.
Handle with caution, this fiery delight,
A force to reckon, with power and might.

Oh Sodium, element so grand,
In your presence, wonders expand.
From ocean depths to fiery blaze,
Your essence forever, in our praise.

SIX

STYLE AND GRACE

In compounds, Sodium hides its face,
A silent partner in a chemical chase.
It bonds with Chlorine, a dazzling affair,
Creating a salt, beyond compare.

With a fiery passion, it seeks to react,
A spark of energy, a volatile act.
Explosive in nature, it's a force to behold,
Handle with care, or danger unfolds.

In laboratories, Sodium finds its home,
A catalyst for progress, where experiments roam.
It dances with elements, creating new wonders,
Unleashing reactions, like rolling thunders.

In nature, Sodium abounds,
In minerals, lakes, and undergrounds.
It colors flames with a vibrant hue,
A vivid display, a spectacle true.

Oh Sodium, element of fame,
In compounds and reactions, you leave your name.
From laboratories to nature's embrace,
You bring your presence, with style and grace.

SEVEN

ESSENTIAL TO LIFE

    Sodium, a metal with a fiery heart
Explosive and reactive, it plays its part
In reactions and compounds, it takes center stage
Its bright yellow flame, igniting like a sage
    A bond with Chlorine, so strong and true
Together they form salt, an essential hue
In laboratories, Sodium acts as a catalyst
Its presence so vibrant, it's hard to miss
    Abundant in nature, it's found all around
In rocks and oceans, it can be found
Its affinity for reactions, a thing to behold
A chemical element, both precious and bold
    So let us celebrate Sodium, with its explosive nature
A metal so unique, with its vibrant feature
A chemical element, so essential to life
Without it, reactions would cease in strife.

EIGHT

SODIUM WHISPERS

In Sodium's realm, a fiery dance takes place,
A catalyst of change, an explosive embrace.
With electrons eager, valence yearning,
Its presence ignites, a fervor returning.

A metal so bright, it dazzles the eye,
In the depths of its flame, secrets lie.
A conductor of currents, a vibrant force,
Sodium, the element, takes its course.

From the salt of the Earth, it does arise,
In oceans and lakes, where beauty lies.
A spark in the night, a flash in the sky,
Sodium whispers, "Behold, I defy."

In compounds it dwells, a partner it seeks,
A taste of the salty, a flavor that speaks.
From soap to streetlights, Sodium plays,
With chemistry's palette, it paints its ways.

So let us ponder this element pure,
A symbol of power, a tale to endure.
In Sodium's realm, mysteries unfold,
A story of fire, a tale yet untold.

NINE

SPARK OF CREATION

In Sodium's fiery heart, a flame does burn,
A catalyst, a spark, a lesson to learn.
With electrons dancing, they leap and collide,
Unleashing reactions, a world open wide.

From nature's embrace, Sodium is born,
In minerals and salts, its presence adorn.
A glowing beacon, a vibrant display,
Its brilliance illuminates the night and the day.

With Chlorine it bonds, a union so strong,
Creating a compound, a bond that's lifelong.
Explosive and reactive, a duo they make,
A force to be reckoned, no boundaries to break.

In oceans and rivers, Sodium resides,
A treasure hidden beneath the tides.
Its essence pervasive, a gift from the sea,
A symbol of life, in all that will be.

With flames that ignite, Sodium reveals,
A tale of its power, its secrets it conceals.
In reactions it dances, a partner so true,
Unleashing its potential, a world to pursue.

Essential to life, Sodium does prevail,
In cells and in bodies, its story we unveil.
A spark of creation, a catalyst unseen,
Bringing harmony to the vast cosmic scene.

So let Sodium's flame forever shine bright,
Igniting our curiosity, our thirst for the light.
In its fiery embrace, we find our own way,
Guided by Sodium, on this journey we stay.

TEN

GRAND DECREE

In the realm of Sodium's glow,
A tale of catalysts we shall sow.
From nature's embrace, it does arise,
A spark of life, a fiery surprise.

In every breath we take, it's found,
In seas and rivers, all around.
A catalyst, it speeds the way,
In reactions, a role it'll play.

From humble Earth, it does emerge,
A symbol of life, an ancient surge.
With Chlorine it dances, hand in hand,
Creating salt, a taste so grand.

Its electrons, so willing to give,
A conductor of currents, it does live.
In fiery flames, it bursts and blazes,
A spectacle of light, as it amazes.

But hidden treasures lie within,
In Sodium's presence, life does begin.
A building block, a vital key,
A part of chemistry's grand decree.

So let us marvel at Sodium's might,
A catalyst that shines so bright.
From nature's embrace, it does unfold,
A symbol of life, in stories untold.

ELEVEN

IGNITING OUR CURIOSITY

In a dance of atoms, Sodium takes its place,
A catalyst of reactions, with fiery grace.
Its electrons, ever eager to shed,
Creating compounds, where reactions are bred.

Abundant in nature, it's never far away,
In oceans and deserts, where it loves to play.
From the depths of the Earth to the sky above,
Sodium's presence is felt, a testament to its love.

A conductor of currents, it sparks with delight,
In the glow of a lamp or a fire's warm light.
With a vibrant explosion, it ignites the scene,
A dazzling display, like a dream unforeseen.

Partnered with Chlorine, Sodium finds its peace,
Creating a union, where salt's flavors increase.

In compounds and reactions, it plays its role,
A building block of chemistry, a vital soul.

Through fiery reactions and vibrant hues,
Sodium guides us on a journey, we can't refuse.
In laboratories and factories, its power we find,
A catalyst for progress, an explorer's mind.

So let us celebrate Sodium's might,
Its affinity for reactions, shining so bright.
Igniting our curiosity, it paves the way,
A guiding element, in the chemistry we sway.

TWELVE

GLORIOUS DREAM

In fiery realms of chemistry's domain,
Where atoms dance and bonds are formed,
A restless element takes its claim,
Sodium, by nature's hand adorned.

With vibrant flame and golden hue,
It casts its light upon the stage,
A catalyst, reactions to pursue,
Unleashing power, with every page.

From sea to sky, in nature's embrace,
Sodium abounds, a versatile force,
In lakes and oceans, its presence we trace,
A creator of compounds, a potent source.

In salts it dwells, with chlorine entwined,
Their union forms a crystal delight,
Enhancing flavors, a taste refined,
Sodium, the seasoning, shining bright.

So let us marvel at sodium's might,
A fiery element, catalyst supreme,
With explosive force and dazzling light,
In nature's realm, a glorious dream.

THIRTEEN

FLASH AND A SIZZLE

In the currents of life, Sodium resides,
A conductor of sparks, where energy collides.
With a single electron, it's eager to share,
Creating reactions, a vibrant affair.

In compounds it dances, a vital ingredient,
Forming salts and solutions, a true achievement.
From table salt to baking soda's might,
Sodium's presence brings flavor and light.

In the depths of the ocean, Sodium lies,
A trace in the water, where life multiplies.
In cells and in blood, its essence is found,
A symbol of vitality, profound.

With a flash and a sizzle, Sodium does gleam,
Its flame burning bright, like a vivid dream.
Partnered with Chlorine, they form a bond,
A union of power, in compounds beyond.

So let us celebrate Sodium's might,
A catalyst of change, a source of delight.
In reactions and compounds, its impact is clear,
Sodium, the element we hold dear.

FOURTEEN

NO HESITATION

In the depths of the ocean's embrace,
Where Sodium dances with grace,
A catalyst of change, it does appear,
With power that is crystal clear.

From the salty waves it does rise,
Forming salts with Chlorine's ties,
Together, they create a bond so strong,
In solutions, their harmony prolongs.

In the quiet currents, Sodium dwells,
In every living organism, it compels,
To spark the fires of life's creation,
A vital element, with no hesitation.

From ancient seas, where it was born,
Sodium's legacy, forever adorn,
In every heartbeat and every breath,
A symbol of life, defying death.

Its electrons, they eagerly share,
With others, in a valiant dare,
To react, to transform, to ignite,
A symphony of chemical delight.
　　So let us sing a song of Sodium's might,
A force of nature, shining bright,
In the depths of the ocean's embrace,
Sodium leaves its indelible trace.

FIFTEEN

CHEMICAL WALTZ

In the realm of elements, Sodium shines,
A catalyst in reactions, its power aligns.
With Chlorine it dances, in a chemical waltz,
Creating salt, a treasure, that never faults.

Sodium, the conductor of an ionic symphony,
It sparks a flame, with electrifying synergy.
Its valence electrons, oh how they collide,
Forming compounds, solutions, where secrets hide.

From the depths of oceans to the earth's crust,
Sodium lies dormant, waiting for a gust.
In nature's pantry, it enhances flavors,
In soups, stews, and all culinary savors.

Its presence, a touch, a subtle sensation,
Heightening tastes with a delicate persuasion.
A pinch of Sodium, a whisper of delight,
Transforming dishes into pure culinary light.

So let us celebrate this element so grand,
A catalyst, a flavor enhancer, at our command.
Sodium, the alchemist's secret ingredient,
We bow to your power, forever magnificent.

SIXTEEN

RARE AND FINE

In the realm of elements, Sodium shines bright,
A catalyst of change, in its atomic might.
Creator of compounds, in reactions it thrives,
Unleashing its power, as nature's alchemic drive.

From the depths of the ocean, Sodium does arise,
In waves and tides, where its essence lies.
It dances with chlorine, in a fiery embrace,
To form the salt that enhances flavors with grace.

In the earth's crust, Sodium we find,
A treasure hidden, of the alchemist's kind.
It weaves its magic, in minerals deep,
A secret ingredient, that its secrets keep.

Oh Sodium, the alchemist's delight,
With your electrons dancing, in the starry night.
You lend your spark, to the flames of creation,
A vital component, in life's grand equation.

So let us celebrate, this element divine,
Sodium, the catalyst, so rare and fine.
In compounds and reactions, it leaves its trace,
A symbol of power, in the cosmic space.

SEVENTEEN

SPEEDING UP PROCESSES

In the realm of elements, Sodium shines bright,
A metal so lively, a brilliant white light.
Its electrons, so eager, they dance and they spin,
Creating reactions, where new bonds begin.

With Chlorine, Sodium forms a grand affair,
A partnership strong, a compound they share.
Together they make salt, that humble condiment,
Enhancing flavors, a culinary sentiment.

But Sodium's story extends far and wide,
In solutions, its presence we cannot hide.
In oceans and rivers, it dissolves with grace,
A crucial component of life's aqueous embrace.

From neurons to muscles, Sodium plays a role,
In transmitting signals, it takes on a toll.

It sparks the synapses, igniting the flame,
Allowing thoughts and actions to take their aim.

Nature, too, holds Sodium in her embrace,
In rocks and minerals, it finds a resting place.
From halite to soda ash, it's found in abundance,
A testament to Sodium's earthly resonance.

A catalyst it is, in chemical reactions,
Speeding up processes, with elegant transactions.
In industries and labs, its power is known,
A driving force, where transformations are sown.

Oh Sodium, element of great might,
With your vibrant presence, the world shines bright.
From compounds to solutions, you leave your trace,
A remarkable element, full of grace.

EIGHTEEN

FIERY ART

In the realm of elements, Sodium I sing,
A metal with a spark, a vibrant thing.
Transmitting signals, sparking synapses,
In the realm of life, where wonder elapses.
Found in nature, in rocks and minerals,
Sodium's presence, a force that endures.
Catalyst in reactions, it plays its part,
Igniting transformations, a fiery art.
Oh, Sodium, with your electrifying hue,
You leave a trace, a memory that's true.
In the depths of oceans, where waves crash and roam,
Your essence dances, creating a vibrant foam.
From fireworks that burst in the night sky,
To the salt on our tables, a taste so spry.
In our bodies, you're a vital component,
Regulating balance, a gift so potent.

So, let us celebrate Sodium's grace,
Its role in the world, we cannot erase.
A chemical marvel, a treasure so bright,
In every atom, a story takes flight.

NINETEEN

SILENT HERO

In the realm of chemistry's grand design,
A star of brilliance doth brightly shine.
Sodium, the catalyst, in reactions bold,
Ignites the fire, a tale yet untold.

Within the crucible of science's domain,
Sodium's presence does not wane.
A catalyst, a spark, a wondrous force,
It sets in motion, a potent discourse.

From the humblest compounds to the grandest flight,
Sodium weaves its magic, casting its light.
In fiery explosions and gentle caress,
Its touch is felt, its power to impress.

A catalyst of change, it speeds the way,
Transforming substances in the heat of the fray.

In chemical reactions, it dances and glows,
Unleashing energy as the story unfolds.
    So, let us raise a toast to Sodium's might,
To its role as a catalyst, shining so bright.
In the alchemy of life, it plays its part,
A silent hero, igniting the heart.

TWENTY

COMPOUNDS AND MIXTURES

In the realm of transformation, Sodium resides,
A catalyst of change, where reactions collide.
With fiery essence, it ignites the night,
A beacon of alchemy, shining so bright.

From the depths of nature, Sodium does emerge,
In oceans and rivers, its presence it purges.
Balancing the forces, it holds the key,
Regulating life's rhythm, in perfect harmony.

A dancer in the elements, it gracefully sways,
In fiery dances, its vibrant essence plays.
With electrons ablaze, it leaps and bounds,
A symphony of reactions, where magic resounds.

In compounds and mixtures, Sodium finds its place,
Creating new substances with elegance and grace.

A catalyst for change, it sparks the unknown,
Unveiling the mysteries, that nature has sown.
    Oh Sodium, mighty element of might,
In the alchemy of life, you shine so bright.
From catalyst to balance, your essence prevails,
In this cosmic dance, where transformation unveils.

TWENTY-ONE

REGULATING BALANCE

In the realm where atoms dance,
Sodium, the catalyst, takes its stance.
With a golden hue, it ignites the flame,
Transforming substances, never the same.

A spark of energy, a catalyst's touch,
Sodium sets reactions in motion, as such.
It speeds the dance of molecules, you see,
Creating new compounds, in perfect harmony.

From the depths of cells to the vast unknown,
Sodium's presence in nature is widely shown.
Regulating balance, it plays a vital part,
Electrolytes flowing, in every beat of the heart.

In oceans and seas, Sodium resides,
Saline waters where life abides.

A symphony of ions, in delicate balance,
Sodium's essence, a cosmic dance.
　So let us marvel at Sodium's might,
A spark of creation, shining so bright.
In the realm of chemistry, it has no peer,
Sodium, the catalyst, forever held dear.

TWENTY-TWO

SALTY WAVES

In the realm of chemistry's embrace,
A catalyst of vibrant grace,
Sodium, the element of flame,
With power and presence, it lays claim.

In reactions, it takes its stance,
Igniting bonds with a fiery dance,
With electrons it shares, it forms anew,
Transforming compounds, old to few.

From the depths of oceans, it does rise,
In salty waves, its essence lies,
A taste of life, a briny sip,
Sodium's touch upon our lips.

In cells and nerves, it plays its part,
Regulating balance, a vital art,
Electrolytes, in harmony they flow,
Sodium's charge, a rhythmic show.

From fireworks to street lamps bright,
Sodium's glow illuminates the night,
A golden hue, a mesmerizing sight,
Guiding our way with its radiant light.

So let us celebrate this element pure,
Sodium's power, forever endure,
In chemistry's tapestry, it weaves its thread,
A catalyst of life, in every spread.

TWENTY-THREE

POLE TO POLE

In the realm of catalysts, Sodium reigns,
A force that stirs reactions, it sustains.
With valiant electrons, it takes its place,
Igniting transformations with its grace.

From metal halides to organic compounds,
Sodium's touch, a symphony resounds.
A spark of life in chemical ballet,
It speeds up the dance, in its own special way.

In nature's grand design, Sodium is found,
In cells and oceans, where life does abound.
It regulates balance, a vital role,
Ensuring harmony, from pole to pole.

With fervent fervor, it seeks to react,
A catalyst of change, that's a proven fact.
From fiery explosions to gentle glows,
Sodium's presence, in every path it shows.

So let us celebrate this noble element,
With its atomic number, so significant.
In the depths of chemistry, it holds its sway,
Sodium, the catalyst, lighting up our way.

TWENTY-FOUR

VIGOR AND MIGHT

In the realm of chemistry's grand ballet,
A dancer emerges, Sodium's display.
A catalyst it plays, in reactions profound,
Creating new compounds, beauty unbound.

Its valence electrons, a single dance,
Eager to lend, to form an alliance.
With elements aplenty, it seeks to combine,
Creating new wonders, a symphony divine.

From salts to soaps, and fireworks bright,
Sodium's touch brings colors to the night.
In water it hisses, with vigor and might,
Releasing its essence, a dazzling sight.

Alkali metal, noble in its stride,
Sodium's presence, the world can't hide.
From oceans to stars, in nature's embrace,
It balances life, with elegance and grace.

TWENTY-FIVE

SYMPHONY OF LIFE

In the realm of chemistry's artistry,
Sodium, a catalyst of great alchemy.
A metal, so bright, it ignites with zeal,
In nature's laboratory, its secrets reveal.

A spark, a flame, Sodium's might,
A catalyst that fuels the chemical fight.
With a gentle touch, it sets reactions ablaze,
Creating new compounds in mysterious ways.

From the salt of the Earth to the depths of the sea,
Sodium dances, as if it were free.
It binds, it connects, it forms endless chains,
Uniting elements in nature's domains.

In the ocean's embrace, Sodium resides,
A beacon of balance, where life coincides.
It regulates rhythms, the body's control,
Ensuring harmony, from head to soul.

With electrons in motion, Sodium unfolds,
A story untold, yet so subtly bold.
In compounds and mixtures, it finds its place,
A vital element, in every living space.

So let us cherish Sodium, this remarkable element,
A catalyst, a regulator, with a purpose so inherent.
In its presence, we find the symphony of life,
A testament to nature's magnificent strife.

TWENTY-SIX

PRESENCE FELT

In the realm of chemistry's dance,
Where elements harmonize and enhance,
There lies a star that shines so bright,
Sodium, the catalyst of boundless might.

From ancient seas, its story begins,
When oceans roared and life had its sins,
It weaved its way through time and space,
Balancing nature's delicate embrace.

Sodium, the conductor of reactions,
Guiding elements with precise actions,
In fiery flames, it bursts to life,
Igniting the darkness with its vibrant light.

A regulator, it controls the tide,
Ensuring harmony, side by side,
From cells to oceans, it does reside,
A presence felt, a force to confide.

With chlorine, it forms a salty bond,
A compound vast, where oceans respond,
And with oxygen, it dances anew,
Creating compounds, both old and true.

Oh Sodium, the element divine,
In chemistry's symphony, you align,
Balancing forces, harmonizing all,
A guiding star, whose light won't fall.

So let us celebrate this shining star,
The catalyst that stretches far,
In every reaction, it plays its part,
Sodium, forever etched in our heart.

TWENTY-SEVEN

THE CATALYST

In chemistry's symphony, Sodium plays its part,
A catalyst of reactions, igniting with a spark.
With its valiant electrons, it dances in the fray,
Creating new compounds in its whimsical way.

From the depths of compounds, Sodium emerges,
A shining light in nature, where balance converges.
In oceans deep, its presence is profound,
Regulating life's rhythm, where harmony is found.

Within the cells, Sodium takes its role,
Maintaining equilibrium, a vital control.
It sparks the signals, a conductor of the nerve,
Guiding the impulses, with a graceful swerve.

Sodium, the catalyst, in reactions it thrives,
A catalyst for change, where chemistry derives.
It brings forth new elements, with a dazzling flare,
Transforming the ordinary into something rare.

Oh Sodium, element of light and fire,
In your presence, reactions never tire.
With your golden glow, you light up the night,
A beacon of potential, shining ever so bright.

So let us celebrate Sodium, this remarkable element,
Balancing forces, with a touch so elegant.
In chemistry's symphony, it takes center stage,
Guiding us on a journey, with wisdom and sage.

TWENTY-EIGHT

SODIUM REVERED

In the realm of reactions, Sodium shines bright,
A catalyst, a force that ignites,
In the crucible of chemistry's art,
It plays its part, a regulator, so smart.

With a spark, it dances, a fiery display,
Unleashing energy in an electrifying way,
In solutions, it dissolves, a willing host,
Guiding reactions, from coast to coast.

From the depths of the ocean to the desert's heat,
Sodium is found, where life and elements meet,
In nature's laboratory, it weaves its spell,
Creating compounds, secrets it does tell.

In the salt of the Earth, Sodium resides,
A taste of the sea, where life abides,
It balances the flavors, a savory touch,
Enhancing dishes, a culinary clutch.

As an electrolyte, it flows through our veins,
Regulating balance, preventing strains,
A vital element, in our bodies it thrives,
Sodium, the guardian of our lives.
   In the laboratory, Sodium is revered,
A symbol of discovery, its presence revered,
From glowing yellow flames to vibrant hues,
It illuminates the path, a beacon it pursues.
   Oh Sodium, remarkable and true,
Guide us through the darkness, we look to you,
With your brilliance and power, you light the way,
In the grand tapestry of elements, you forever stay.

TWENTY-NINE

CATALYST SUPREME

In the realm of chemistry, a star does reside,
A catalyst of change, with power to guide.
Its name is Sodium, a shimmering light,
A force that ignites, with sparks burning bright.

In reactions it dances, a regulator so fine,
Balancing the elements, like a conductor in time.
It binds with chlorine, a marriage so pure,
Creating a compound that will forever endure.

An element of wonder, Sodium does bring,
A taste of the ocean, a song that does sing.
In waves it dissolves, in the depths it resides,
A treasure beneath, where life's secrets hide.

From the depths it emerges, in crystals so white,
A spark in the darkness, a beacon of light.
It dances with fire, a flame that does soar,
A symphony of colors, like never before.

Oh Sodium, element of grace and might,
You guide the reactions, like day turns to night.
In the lab, you're a hero, a catalyst supreme,
With your presence, miracles can be seen.

So let us celebrate, this element divine,
For Sodium's presence, forever we'll pine.
A catalyst of change, a regulator so grand,
In the world of chemistry, you shall forever stand.

THIRTY

EVER TRUE

In the realm of elements, Sodium takes its hold,
A catalyst, a regulator, its story to be told.
A conductor in nature's grand symphony,
Bringing balance, light, and transformation with glee.
In culinary realms, Sodium dances with delight,
Enhancing flavors, a chef's guiding light.
From the humble salt that seasons our food,
To the crispy delight of pretzels, so good.
Within our bodies, Sodium plays its part,
Regulating fluids, keeping rhythm in our heart.
Electrolyte balance, a delicate dance,
Ensuring vitality, giving life a chance.
In science's domain, Sodium takes flight,
A fiery explosion, a dazzling sight.
When dropped in water, a reaction so pure,
Bursts of energy, a spectacle to endure.

From stars in the cosmos to the depths of the sea,
Sodium's presence is felt, for all to see.
A symbol of transformation, a beacon of grace,
In every aspect, Sodium leaves its trace.

So let us celebrate this element of might,
Sodium, a force that brings us light.
In every realm, it guides us through,
A reminder of nature's wonders, ever true.

THIRTY-ONE

NEW WORLDS EMERGE

In the depths of the ocean, Sodium does reside,
A shimmering presence, a force to confide.
It dances with waves, a celestial ballet,
Regulating life in a mystical way.

With ions in tow, it traverses the sea,
Balancing forces, in perfect harmony.
From muscle contractions to nerve impulses,
Sodium's touch, a conductor that never ceases.

In culinary arts, it adds flavor and zest,
Enhancing each dish, the very best.
From table salt to baking soda's might,
Sodium's touch, a culinary delight.

In our bodies, it plays a vital role,
Maintaining balance, it takes control.

Fluid levels regulated, blood pressure kept,
Sodium's presence, a secret well-kept.

A catalyst it is, in reactions untold,
Creating compounds, a treasure to behold.
From Sodium chloride to Sodium hydroxide,
New worlds emerge, with Sodium as guide.

In the darkest of nights, it illuminates the path,
A beacon of light, dispelling the wrath.
Sodium's flame, a mesmerizing sight,
Guiding us forward, with its radiant might.

Oh, Sodium, an element so profound,
In the realm of chemistry, you astound.
From oceans to bodies, from reactions to flame,
You are a remarkable element, Sodium by name.

THIRTY-TWO

ENHANCING THE TASTE

In the realm of elements, Sodium stands tall,
A guardian of balance, in bodies, it sprawls.
Electrolyte supreme, in cells it resides,
Regulating functions, where harmony abides.

In reactions it dances, with fervor and might,
A catalyst of change, in the chemical fight.
With chlorine it bonds, as Sodium Chloride,
A union so strong, that oceans can't hide.

In nature it gleams, in the fiery sun,
A star's shining essence, where fusion is spun.
In salts and in minerals, Sodium is found,
A treasure of Earth, in abundance profound.

But beyond its presence, in science and lore,
Sodium weaves tales, of flavors galore.

In kitchens it thrives, as a seasoning grand,
Enhancing the taste, with a skilled chef's hand.
 From pickles to pretzels, and breads that rise,
Sodium brings joy, to our taste buds' surprise.
A pinch of its magic, can transform a dish,
A symphony of flavors, in a single wish.
 So let us celebrate, this element divine,
Sodium, the alchemist, in every design.
From the tiniest cells, to the grandest of feasts,
It guides us through life, as nature's own priest.

THIRTY-THREE

REIGNS SUPREME

In the depths of our bodies, Sodium resides,
A humble element, with secrets to confide.
It dances in the fluids, flowing through our veins,
Regulating balance, where harmony remains.

A guardian of cells, a conductor of charge,
Sodium, the element, large and so vast,
It sparks the synapses, ignites the mind's fire,
Guiding impulses, like a string in a lyre.

Sodium, the alchemist, in reactions it thrives,
Unleashing its power, where chemistry derives,
A catalyst of change, a builder of bonds,
Creating compounds, where new life responds.

From saline oceans to crystalline shores,
Sodium, the seasoning, enhances flavors galore,
In culinary arts, it adds a savory touch,
A pinch of perfection, a taste loved so much.

And in the Earth's abundance, Sodium prevails,
A treasure of nature, where its essence unveils,
From table to laboratory, it serves us with grace,
Sodium, the element, an integral part of our space.

So let us celebrate this element divine,
For Sodium's presence, in us, will always shine,
In balance and flavor, in reactions unseen,
Sodium, the element, forever reigns supreme.

THIRTY-FOUR

WORLD ADORNS

In the depths of the Earth, Sodium resides,
A shimmering metal, where secrets reside.
A beacon of light, in the darkness it gleams,
Unveiling the wonders, of life's endless themes.

Within our bodies, it dances and plays,
Regulating rhythms, in marvelous ways.
Electrolyte hero, it keeps us in sync,
Balancing fluids, so we never sink.

In the laboratory, Sodium takes flight,
Reacting with brilliance, igniting the night.
A catalyst bold, forging compounds anew,
Unleashing reactions, for science to pursue.

On the culinary stage, Sodium performs,
A pinch of its magic, enhances all norms.
From humble salt shakers to sizzling pans,
It seasons our meals, with skilled artistry's hands.

In nature's embrace, Sodium finds its place,
In oceans and rivers, it leaves its trace.
From salts and minerals, it weaves a grand tale,
Of life's evolution, from sea to land's vale.

Oh, Sodium! Element of light and might,
Guiding us onward, through darkness and light.
With balance and flavor, you grace our existence,
A symbol of progress, and life's persistence.

So, let us celebrate, this element so bright,
With poems and songs, and joyous delight.
For Sodium's essence, in all its grand forms,
Embodies the wonders, that this world adorns.

THIRTY-FIVE

DIVERSE AND GRAND

In the realm of atoms and the periodic table,
There exists a shining element, Sodium, stable.
With an atomic number of eleven, it stands tall,
A metal so bright, its presence enthralls.

Within our bodies, Sodium does reside,
Regulating fluids, maintaining balance inside.
In every cell, it plays a vital role,
Ensuring hydration, keeping us whole.

From the oceans deep, Sodium does emerge,
Dissolved in seawater, a salty surge.
It dances with Chlorine, forming a bond,
Creating salt, a flavor beyond.

In the flames of fire, Sodium ignites,
A brilliant yellow glow, a captivating sight.
A catalyst it becomes, in reactions it thrives,
Speeding up the process, where chemistry thrives.

Sodium, a culinary delight,
Enhancing flavors, a chef's true might.
From savory dishes to sweet delights,
Its presence in recipes, a true gastronomic height.

In nature's grand design, Sodium has a place,
In the evolution of life, it left its trace.
From ancient oceans to creatures of the land,
Sodium's influence, a guiding hand.

So let us celebrate Sodium's might,
In every aspect, it shines so bright.
A versatile element, diverse and grand,
Sodium, the essence, of life's intricate strand.

THIRTY-SIX

CLOSE TO OUR HEART

Sodium, a metal of great repute,
In culinary arts, it's an absolute brute.
A dash of salt, a pinch of spice,
Sodium's presence makes dishes nice.

But outside the kitchen, it has a role,
In nature's evolution, it played a vital role.
From the oceans to the land,
Sodium's abundance shaped the grand.

It's not just a taste enhancer,
But a chemical element with an answer.
In chemistry labs, it's a key,
Reacting with others to set reactions free.

And in the night sky, it sparkles bright,
As fireworks explode with sodium's might.
Orange-yellow flames, a sight to see,
Sodium's beauty, a wonder to be.

But in bread-making, it's truly divine,
Sodium carbonate, the secret behind.
Fluffy loaves with a crispy crust,
Sodium's influence, we cannot distrust.

Oh, Sodium, a metal so fine,
Enhancing life and flavors divine.
In every aspect, it plays a part,
A chemical element, close to our heart.

ABOUT THE AUTHOR

Walter the Educator is one of the pseudonyms for Walter Anderson. Formally educated in Chemistry, Business, and Education, he is an educator, an author, a diverse entrepreneur, and he is the son of a disabled war veteran. "Walter the Educator" shares his time between educating and creating. He holds interests and owns several creative projects that entertain, enlighten, enhance, and educate, hoping to inspire and motivate you.

Follow, find new works, and stay up to date
with Walter the Educator™
at WaltertheEducator.com

www.ingramcontent.com/pod-product-compliance
Lightning Source LLC
LaVergne TN
LVHW020134080526
838201LV00119B/3770